iScience
Readers

Food and Nutrition:
Eating to Win

by Emily Sohn and Diane Bair

Chief Content Consultant
Edward Rock
Associate Executive Director, National Science Teachers Association

NORWOOD HOUSE PRESS
Chicago, Illinois

Norwood House Press
PO Box 316598
Chicago, IL 60631

For information regarding Norwood House Press, please visit our website at
www.norwoodhousepress.com or call 866-565-2900.

Special thanks to: Robert Collison, MS, RD, CDE; Amanda Jones, Amy Karasick,
Alanna Mertens, Terrence Young, Jr.

Editors: Michelle Parsons, Barbara J. Foster, Diane Hinckley
Designer: Daniel M. Greene
Production Management: Victory Productions, Inc.

Paperback ISBN: 978-1-60357-290-3

The Library of Congress has cataloged the original hardcover edition with the following
call number: 2010044545

Printed in Heshan City, Guangdong, China.
190P—082011.

CONTENTS

Note to Caregivers:

Throughout this book, many questions are posed to the reader. Some are open-ended and ask what the reader thinks. Discuss these questions with your child and guide him or her in thinking through the possible answers and outcomes. There are also questions posed which have a specific answer. Encourage your child to read through the text to determine the correct answer. Most importantly, encourage answers grounded in reality while also allowing imaginations to soar. Information to help support you as you share the book with your child is provided in the back in the **Additional Notes** section.

Words that are **bolded** are defined in the glossary in the back of the book.

The Power of Food

Bananas, cookies, carrots, pizza, potato chips, broccoli. Life is full of things to eat. And eating is something most of us do at least three times a day. But how do you choose what foods to include in your diet? You might not think much about it. But you should. Your food choices make a big difference in your life. Eating well can make you stronger, happier, and smarter. In this book you will learn about power foods that are beneficial for good health. You will also learn about foods that take your power away. You will use what you learn to help runners finish a marathon, with energy to spare.

Fueling Up

You own a restaurant. It lies right at the start of a popular running race. The race is tomorrow. But it's not just any race. It is a marathon. Starting at 8 a.m. tomorrow, the runners will cover 26.2 hilly miles (42.2 kilometers). In a car, it would take half an hour to drive that far on a fast highway. On foot, it takes hours.

Food is often involved when people get together for almost any reason.

Thousands of people are traveling to your city to run in the race. Many of them are coming to eat at your restaurant tonight. The runners have been training for months. Their bodies are in top condition. Dinner is the last big meal they will eat before the big day. What will you serve them to keep them running strong tomorrow? Here are the choices.

Meal 1:

A huge salad full of vegetables and fruit, a chocolate cookie, and a big glass of water.

Meal 2:

A hamburger with lettuce, tomato, and ketchup; a handful of almonds; and a large soda.

Meal 3:

A big plate of pasta with tomato sauce, shredded cheese, a small salad with walnuts, and a big glass of low-fat milk.

Meal 4:

Fried fish sticks with tartar sauce, potato chips, and a blue sports drink.

Planning a Healthy Diet

It's not a good idea to always eat just one thing, even if it's a healthy food. You will miss some important other **nutrients.** Eating a wide variety of foods is a better plan. That way, you are much more likely to get everything you need to be healthy, fast, and strong.

Look at the Food Guide Pyramid. It sorts foods into groups. It also shows what, and how much, to eat from each food group. You need to eat foods from every group. You should get more food from the groups that are in the biggest triangles. What does the pyramid tell you about oils?

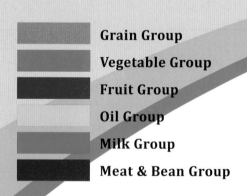

Grain Group

Vegetable Group

Fruit Group

Oil Group

Milk Group

Meat & Bean Group

MyPyramid.gov
STEPS TO A HEALTHIER YOU

Do you eat foods from each group every day? How can the food pyramid help you plan a training menu for the marathon runners?

One good tip is to "eat the rainbow." Pick foods that burst with color, such as red peppers, orange carrots, and dark, leafy greens. Colorful fruits and vegetables like these are packed with **vitamins** and **minerals.** Fruits and vegetables also contain a lot of **fiber,** which is necessary for healthy digestion.

A meal that has a healthful mix of nutrients, such as this one, is called a balanced meal.

A healthy meal contains a balance of nutrients. It draws from the whole pyramid. But it doesn't have too many foods from any one section of the pyramid. The dinner in the photograph has a perfect mix of lean **proteins,** complex **carbohydrates,** and good **fats.** You will read more about these nutrients starting on page 14. The meal also contains good amounts of fiber, vitamins, and minerals.

Plan a restaurant menu for training during the week leading up to the marathon. Show how the runners who come to your restaurant will "eat the pyramid." Which of the choices in the puzzle have the most colors? Which have the fewest?

Why Do You Need to Eat?

Grumble, grumble: When your stomach talks, it can be hard to ignore. Hunger is a powerful feeling. But why do we need to eat? One reason is that food gives us energy. All the energy we gain from food begins in the Sun.

Corn is a plant food. Corn gets its energy directly from the Sun.

Most plants capture energy from sunlight and make their own food. They store this energy in their cells. Cells are basic building blocks of all living things. Plant foods, like carrots or kale, give you some of their energy when you eat them. You also get some of the Sun's power when you eat meat or eggs. These products come from animals that eat plants.

To get energy out of your food, the cells in your body do something called cellular **respiration.** In this process, cells turn the nutrients in our food into the **molecules,** or chemicals, that our bodies use as fuel. What kinds of animals do people eat? What do those animals eat?

Look at the choices in the puzzle. Which foods get their energy directly from the Sun?

Vigorous physical activity requires lots of energy, which means it uses lots of calories.

Too Much, Too Little, Just Right

Some foods pack more power than others. You can tell how much energy a food contains by finding out how many **calories** are in it. An apple has about 100 calories. A cup of lettuce has fewer than 10 calories. And a half-cup of peanuts has more than 400 calories. The more calories a food has, the more energy it carries.

The number of calories you need depends on your size, your age, and whether you are a boy or a girl. If you are really active, you need more food than someone who is your size but doesn't run around as much. The more you move, the more energy you use up.

Marathon runners can burn, or use the energy from, more than 2,000 calories during a race. Which choices in your restaurant puzzle offer the most calories? Which do you think offer the fewest calories? Who do you think will need more calories, a big runner or a small runner? Explain your answer.

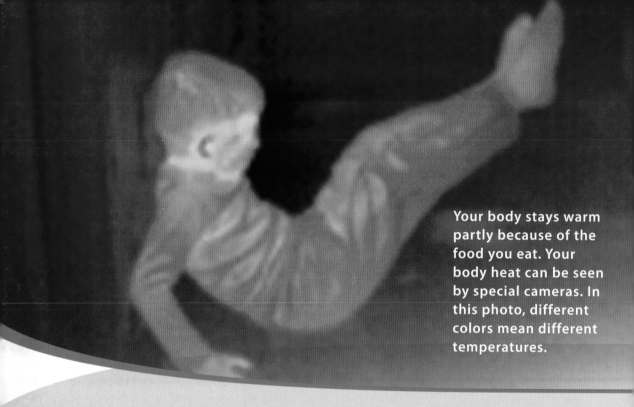

Your body stays warm partly because of the food you eat. Your body heat can be seen by special cameras. In this photo, different colors mean different temperatures.

Runners are like machines. Food is like the batteries that make them go. We all need a certain amount of energy just to breathe and walk around. The process of turning food to energy is called **metabolism.** When your body turns fuel into energy, it also creates heat. That's why you get hot and sweaty when you run really fast. Everybody's metabolism is different. Your body needs a certain amount of calories to do what you need to do.

But there is a balance. If you eat many more calories than you use up, you will gain weight. If you eat too few calories, you lose weight and run low on energy. Eating too much or too little keeps you from doing your best. The healthiest way to live is to eat just enough.

What do you think would happen if a marathon runner ate too few calories while training? What if the same runner ate too many calories? Do you think runners need to eat more than or less than normal the night before a race? Do you think calories affect how fast and how far these athletes can run?

Why Do You Need to "Eat the Pyramid"?

You want your restaurant to be popular with the runners. So, you had better set a good example. As you plan your menu for the athletes, look at your own diet. For an entire day, write down everything you eat.

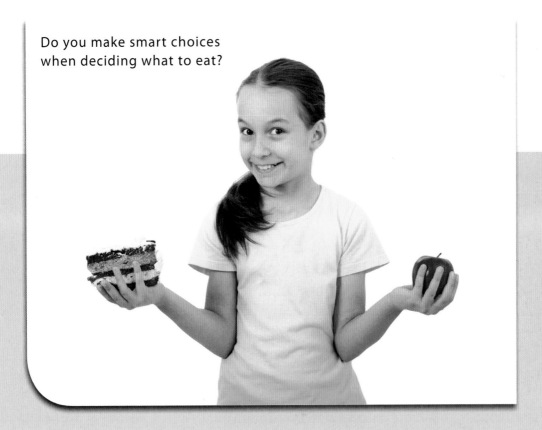

Do you make smart choices when deciding what to eat?

Are you getting enough food from all the food groups to stay healthy? It can be hard to tell. By choosing foods carefully, you can get everything you need in the right amounts. Do you think you choose foods carefully?

Read on to learn more about which foods have more vitamins, minerals, and other things your body needs. See if you can eat like a marathon runner for a day.

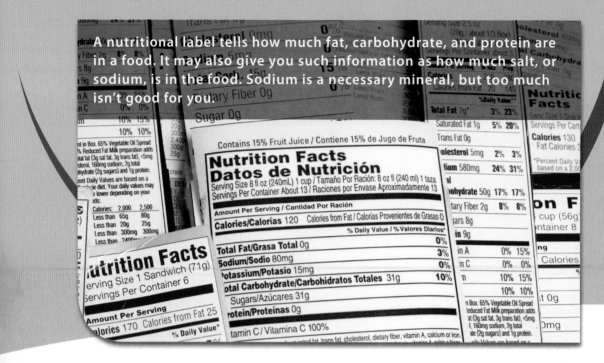

A nutritional label tells how much fat, carbohydrate, and protein are in a food. It may also give you such information as how much salt, or sodium, is in the food. Sodium is a necessary mineral, but too much isn't good for you.

Three Kinds of Power

All foods give us energy. But they don't all deliver energy in the same way. Food falls into three main groups. These groups are called carbohydrates, proteins, and fats. Each offers energy in its own way. They are all part of good **nutrition.**

(Note: The Food Guide Pyramid shows six groups. Still, in terms of nutrition, the foods in each group can be labeled carbohydrate, protein, or fat.)

Foods that are high in carbohydrates include bread, pasta, potatoes, and fruit. High-protein foods include beans, meat, milk, eggs, and fish. High-fat foods include butter, oil, and nuts. Fats are found in foods that come from both animal and plant sources.

We'll talk more later about carbohydrates, proteins, and fats in our daily diet. For now, just know that we need all three. We also need vitamins and minerals to stay healthy.

Look at the menu options in the restaurant puzzle. Which do you think has the most protein? Which has the most fat? Which has the most carbohydrates?

The Quick Fix

The night before a race, athletes often eat lots of carbohydrate. This practice is called "carbo-loading." It's a good plan because carbohydrates are the body's major energy source. The body turns to them first for quick energy. There are two kinds of carbohydrates in food: starches and sugars. Starches provide energy for a longer period of time than sugars do. You get starches when you eat bread, pasta, potatoes, grains, and beans. These foods take a while to digest.

Some carbohydrates give you quick energy. Other carbohydrates give you long-lasting energy.

Sugars are a more simple form of carbohydrates. You get natural sugars from fruit, milk, and honey. Candy and cookies have a lot of sugar added to them. Eating simple sugars can lead to a quick burst of energy. But the energy burst doesn't last as long as the energy you get from eating starches. You should get about 45 to 65 percent of your calories each day from carbohydrates. That's about half of your calories.

On the evening before the marathon race, the runners who come to eat at your restaurant will need an extra dose of carbohydrates. A little protein and fat are good. But too much can weigh a runner down right before a race. Look at the puzzle answers. Which choice offers the most starch? Which choice offers the most sugar? Do you think the runners will want more starch or more sugar before the race? Explain your answers.

People all over the world eat whole foods, such as beans, nuts, and dried fruits. They are much healthier than processed foods.

The best carbohydrates are what nutrition experts call "whole foods." Some, such as beans, vegetables, and fruit, come right off a plant. These are called **unprocessed** foods. Simple foods with just a few ingredients are also good choices. Examples include breads and cereals made from whole grains. It doesn't take a lot of work in a factory to make whole foods. Most whole foods have a lot of healthy fiber.

Do you know why candy and other sugary treats are called "empty" calories? Do any of your menu choices contain **processed** foods? Would these foods help a runner during a race?

Sports drinks and sodas contain lots of sugar. Use them only as an occasional treat. Water is a better choice most of the time.

Sugar shows up in places you might expect, such as in soda pop, candy, and cake. Sugar also lurks in places that might surprise you. There is sugar in fruit juice, ketchup, and some kinds of peanut butter.

Start reading labels on your favorite snacks and drinks. To find out how much sugar you're getting, first look at the number of grams of sugar in each serving. Multiply this number by 0.2. The result tells you how many teaspoons of sugar you just ate. Five grams of sugar equals one teaspoon ($5 \times 0.2 = 1.0$). Use this formula to find out how many teaspoons of sugar you get each day just from what you drink. It is easy to get way too much sugar. Are you drinking more sugar or less sugar than you thought?

Every meal on your restaurant menu offers a drink. Which drink do you think has the most sugar? Which are the best drinks for the runners? Which are the worst drinks for them?

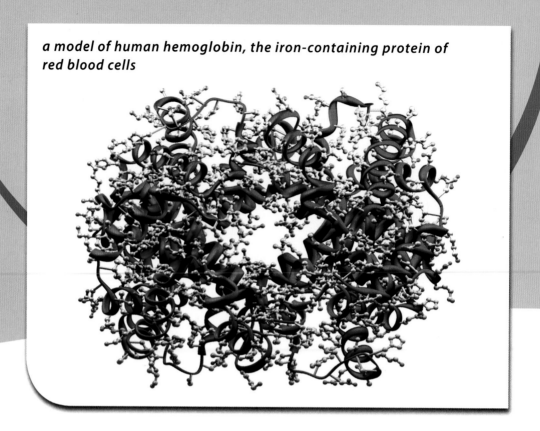

a model of human hemoglobin, the iron-containing protein of red blood cells

Eating for Power

It is important to eat carbohydrates the night before a race. But eating to train for a race is a different story. Proteins keep our bodies strong, day after day and month after month. They help build body tissues. They appear in skin, hair, blood, bones, and fingernails. Proteins help repair injured tissues. And they help build strong muscles. That's where running power begins.

Which items on your menu are best for runners who come to your restaurant a few weeks before the race? Would you offer runners something different a week before than you would a day before the race? What would you offer them for lunch the day after the race?

Some vegetarians avoid eating meat. Others are stricter, and they avoid eating any animal products, including milk, cheese, and eggs.

Proteins are made up of 20 building blocks called **amino acids.** Not all proteins contain the same combination of these molecules. Meat, milk, eggs, and cheese have complete proteins. Complete proteins have all the amino acids we need to build muscle and other tissues. These foods all come from animals.

Some plant foods, such as beans, avocados, and a grain called quinoa, also contain proteins. Some plant foods have **complete proteins.** Other plant foods have **incomplete proteins.** When eating incomplete proteins, you have to combine them in your diet to make them complete. Eating rice and beans in the same meal will do the trick.

Athletes don't have to eat meat to run fast. People who don't eat meat are called vegetarians. Some vegetarians don't eat any foods from animals at all. Which choices on your menu are vegetarian?

A serving of meat should be about the size of a deck of cards. That gives you just the right amount of protein in your diet. How much meat do you eat in one meal?

Even though protein is an important nutrient, you do not need to eat a lot of it to stay healthy. Your body actually makes some amino acids on its own. That means that you do not have to get all the amino acids you need from your diet. Ten to fifteen percent of your daily calories should come from protein. Some animal sources of protein include chicken, fish, beef, milk, cheese, and eggs. Plant sources include nuts, beans, peanut butter, and some vegetables.

Which meal choices in the puzzle contain the right amount of protein? Which have too much? Which have too little?

Eating for the Long Haul

After carbohydrates and proteins come fats, the third main group of foods. Some people worry that eating fats will make them fat. But fats are important for your body. They fuel your brain. They help your body absorb vitamins. And they protect your organs, such as your liver and your heart. Runners need fats to help them go extra-long distances. At the beginning of a race, athletes burn carbohydrates. When the carbs run out, their bodies start burning fat. Fat provides energy over the long haul.

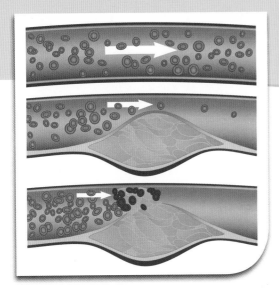

a healthy artery

a partially blocked artery

a completely blocked artery

Fats come in a variety of forms. Some of the most healthful fats appear in nuts, olive oil, and fish. Salmon is one of the healthiest kinds of fish. Not-so-good fats come in meat, cheese, and junk food, such as chips, candy, and cookies. These fats are sometimes called "bad fats." It's okay to eat them in small amounts. But too much fat can plug up your arteries. Arteries are the tubes that carry blood away from your heart and to the rest of your body. With clogged arteries, your heart has to work harder than it should have to. That can lead to heart disease.

Which items on your menu have good fats? Which items have bad fats?

21

Fats from olives and olive products are good for you in small amounts.

Your diet is like a seesaw. You want to find just the right balance, without eating too much or too little of any one thing. It is easy to overdo it on fats, because fats pack more calories per serving than proteins and carbohydrates do. Try to eat good fats in small amounts.

One healthful strategy is to go for low-fat versions of chicken, milk, and cheese. Foods cooked in olive oil are healthier than foods cooked in butter. Grilling and baking are healthier than frying. Nuts are better snacks than chips.

What kinds of fats are in the foods you usually eat? How could you get rid of some of the unhealthy fats from your diet?

Some fats, such as those in olive oil, are better for your health than other fats, such as those in butter. Still, you should eat them in small amounts.

Why Do You Need Vitamins and Minerals?

You might take vitamins that come out of a jar. But what's inside those little nuggets? Vitamins are nutrients that you need in small amounts. Vitamins do a lot of things. They help fight infections. They keep your organs functioning. And they help you learn. Your body is a powerful machine, but it cannot make most vitamins. Just about all the vitamins you need come from the foods you eat. Here are a few ways vitamins can help you:

Vitamin A	carrots sweet potatoes cantaloupe	Good for vision
B Vitamins (a group of vitamins)	whole grains lentils molasses	Help your metabolism and digestion
Vitamin C	lemons tomatoes red peppers	Good for healing wounds and burns
Vitamin D	salmon milk tuna	Builds strong bones and teeth
Vitamin K	broccoli kale spinach	Helps your blood clot when you get a cut
Vitamin E	almonds sunflower seeds peanut butter	Good for red blood cells and muscle tissue

Think about the marathon runners who will be coming to your restaurant tonight. Which vitamins should they make sure they're getting enough of? Use the chart to check the vitamin content of your menu choices.

Choose mineral-rich foods for healthy blood as well as strong bones and teeth.

Along with vitamins, you need minerals in small amounts, too. Calcium and potassium are well-known minerals. Iron, magnesium, and zinc are other minerals. We need to get about 15 different minerals from the foods we eat. The best way to get them is to eat a variety of fresh and colorful foods.

Minerals have a variety of jobs. Calcium helps build strong bones and teeth. Milk and cheese are rich in calcium. Potassium can fight cramps during exercise. It is found in bananas and leafy greens. Iron is good for the blood. It appears in meat, leafy greens, beans, and many breakfast cereals. All of these minerals help runners perform better.

Now, look at your menu again. Which choices do you think have the most minerals?

Filling the Tank

Food is the star of a restaurant menu. But the liquids you drink are just as important as what's on your plate. Water is most important of all. People can survive for only a few days without drinking water. We can go much longer without food, though.

Water is the most important thing you can put into your body.

The human body is about two-thirds water. We lose water through sweat when we exercise and when the temperature is high. Simply breathing and walking around also use up water. If the water level in your body dips too low, you become dehydrated. Illnesses can also cause **dehydration.** People who are dehydrated may feel dizzy. They can have fevers or headaches. Sometimes, they even pass out.

One way to "fill your tank" is to drink water throughout the day. Drinking juice and other liquids helps, too. You can also get some liquid out of fruit, vegetables, and other juicy foods.

The runners will be sweating a lot during tomorrow's race. How can you help them hydrate beforehand? How do you think the runners should stay hydrated during the race?

Vitamins, minerals, and other important nutrients are all tied up in your food. So, how does your body get those nutrients out of the food you eat? Your digestive system is the factory that turns food into fuel. The process starts every time you open your mouth and take a bite.

Too Big to Eat

Have you ever seen a hot dog eating contest? People have a few minutes to eat as many hot dogs as they can. The winners barely chew their food. It can be funny to watch. But eating that way is a bad idea. Chewing breaks our food into smaller pieces. That gives the digestive system something to work with. The digestive system pulls nutrients out of your chewed food. Your cells and tissues can then take up the nutrients and use them.

Your digestive system has to do a lot of extra work if you don't use your teeth and saliva to start breaking down food before you swallow it.

How long did it take you to eat breakfast this morning? Try eating more slowly at your next meal. Write down how you feel after you eat quickly or slowly. Do you think the marathon runners should eat their dinners quickly or slowly?

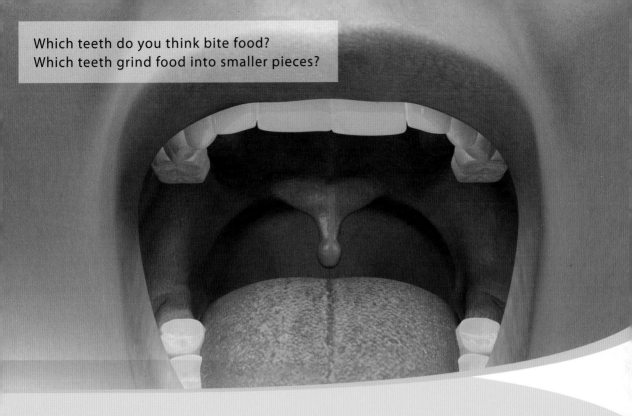

Our mouths are really good at breaking down the food we eat. Your tongue and teeth work together to turn big chunks of food into smaller pieces that are easier to swallow. Why do you think we have sharp, thin teeth in the front and wide, flat teeth in the back?

Saliva helps with digestion, too. It doesn't break food apart physically like teeth do. Instead, it does something chemical. It starts the process of turning a solid food into its nutrient parts. That's right: Saliva is good for more than just spitballs!

27

Your digestive system works something like this wood chipper.

Have you ever seen a landscaper using a wood chipper? The landscaper takes giant pieces of tree limbs and puts them into a chipping machine. The chipper grinds the big chunks of wood into much smaller pieces. Those tiny bits of wood can be spread around a garden as mulch, to improve the soil and help create a healthier garden.

Your digestive system is something like that. It breaks down the food you eat into molecules that your body can use. Do you know which parts of the body play a role in digestion?

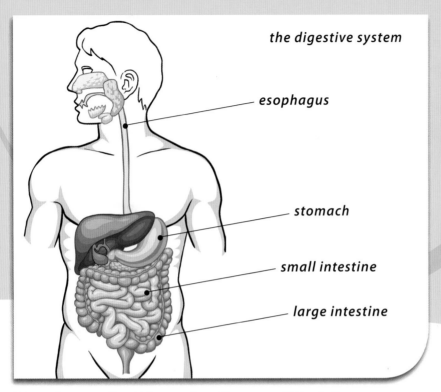

the digestive system

esophagus

stomach

small intestine

large intestine

How Does the Digestive System Work?

Imagine one of your runners sitting down for breakfast the morning of the race. On the menu: Toast with peanut butter and banana, and juice. With the very first bite, the bread begins its journey through the runner's body.

First, her teeth break the food apart. Her molars grind the crust, peanut butter, and fruit into smaller pieces. Saliva softens the food and begins the chemical breakdown. Her tongue helps push the mush into her throat. Now, she can swallow.

The food's first destination is the **esophagus.** This is a tube that connects the throat to the stomach. Muscles in the esophagus squeeze to push food to the stomach.

Think about your dinner menu. Which menu items do you think would be easiest to chew? Which foods would probably be harder to grind down?

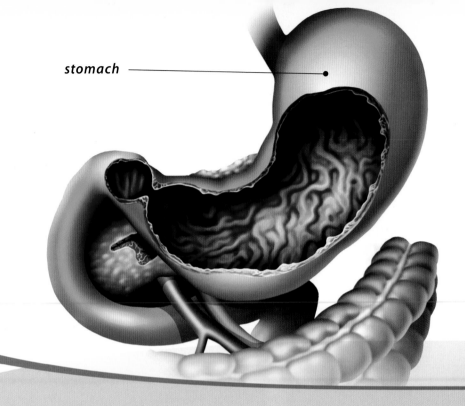

stomach

The stomach looks like a sac in the shape of the letter J. It has strong muscles that churn the food into small pieces. Stomachs also have chemicals called acids and **enzymes.** These molecules turn chewed-up food into even mushier stuff. Food usually stays in the stomach for about two hours.

Acids are strong chemicals. A thick layer of **mucus** on the inside of the stomach keeps these chemical juices from eating away at your stomach.

It can be uncomfortable to run with a lot of food in your stomach. The race starts at 8 a.m. tomorrow. What time do you think the runners should eat breakfast?

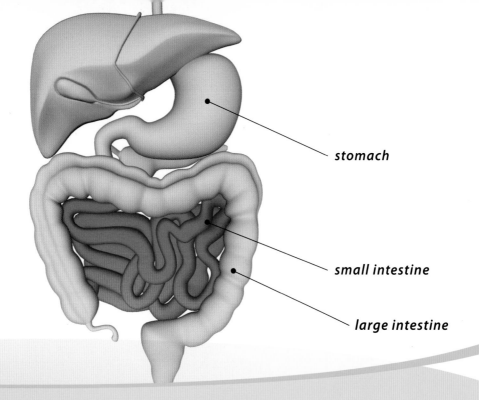

stomach

small intestine

large intestine

Bang! Imagine the starting gun has gone off. The runner is on her way. She finished eating breakfast long ago. But her breakfast is still winding its way through her digestive system. Next stop: The small intestine. This is a coiled tube that lies behind the belly button. If you stretched the organ out, it would be 20 feet (6.1 meters) long. In the small intestine, proteins, carbohydrates, fats, vitamins, and minerals get into the bloodstream. From there they travel to all the cells in the body.

Some food does not get digested. This waste goes into the large intestine. There, any water that is mixed in with the food is taken out. Solid waste remains. Both liquid and solid wastes leave the body when you use the bathroom.

Marathoners run for hours. Do you think they want to eat food that is easy or hard to digest? Why?

Fruits have a lot of fiber, which is necessary for digestion.

Just as you can keep a machine in good repair, you can help your digestive system work better. One way is to drink a lot of water and eat a healthy diet. Choose foods with a lot of fiber, such as fruits, vegetables, and whole grains. Fiber makes it easier for food to pass through your system.

Continue to record everything you eat and drink for one week. Also write down how much energy you feel you have throughout the day. How hungry are you at different times? Which foods give you the most energy for the longest time? Use this information to help you plan your training menu for the runners.

High divers, runners, and sumo wrestlers have different training needs and different nutritional needs. A good coach knows what is right for a particular athlete.

SCIENCE AT WORK

Coaches

Athletes need to learn how to train without getting hurt. Coaches teach athletes how to get better at their sports. They teach a runner about body position and proper breathing techniques, for example. They make sure their athletes warm up, cool down, and stretch appropriately. They also help athletes learn how to fuel their bodies to reduce the chance of injury. They teach athletes what to eat to perform as well as possible.

Different athletes have different needs. Think of a high-diver, a runner, and a sumo wrestler. A good coach will know an athlete's nutritional needs, based on body type, weight, kind and amount of activity, and so on. The coach will tell an athlete how much carbohydrate, protein, and fat to eat. A coach will also tell the athlete when to eat, and what foods to eat before an event and after an event.

How Do Other Body Systems Help Digestion?

Digestion is a huge job. The job is so huge, in fact, that one group of organs can't do it on its own. To break down everything you eat and drink, your digestive system gets help from other parts of the body.

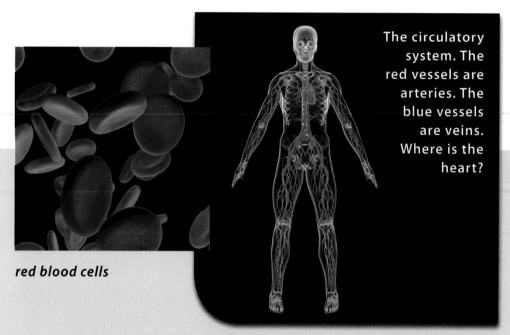

The circulatory system. The red vessels are arteries. The blue vessels are veins. Where is the heart?

red blood cells

Blood to the Rescue

One of the biggest helpers is the **circulatory system.** This group of parts includes the heart, blood vessels, and blood. The heart pumps blood, which travels around the body. Blood moves through the blood vessels, also called arteries and veins. Blood carries nutrients and oxygen to the cells. It also carries wastes away.

Some scientists estimate that we have about 100 trillion cells in our bodies. That is a 1 followed by 14 zeros—100,000,000,000,000. Those are a lot of cells to feed!

When marathoners run, their hearts work hard to pump a lot of blood through their bodies. Do you remember which mineral is good for the blood? Which foods contain it?

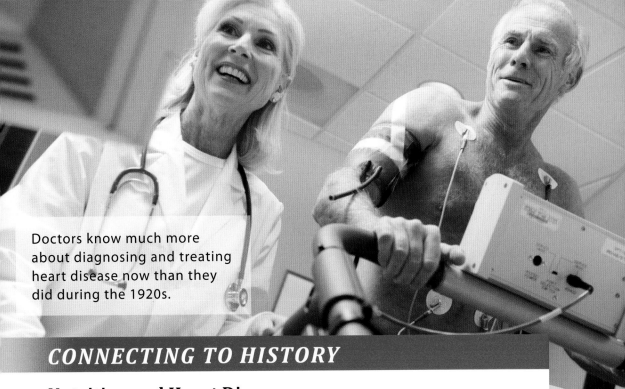

Doctors know much more about diagnosing and treating heart disease now than they did during the 1920s.

Nutrition and Heart Disease

During the 1920s, heart disease became the most common cause of death in the United States. It remains so today. At first, scientists didn't know why. Something must have changed to put Americans at higher risk for heart disease. But what?

Medical researchers went to work. They compared healthy people to those with heart disease. They tried to figure out how people's lives had changed. They wanted to know if it was something people did or did not do every day that made them more likely to get heart disease.

Imagine you are a researcher. You have no idea why a disease is suddenly more common than it used to be. What questions would you ask people to help solve the mystery?

When more people started driving cars and eating processed foods at fast-food restaurants, instead of walking and eating healthful meals, heart disease became much more common.

The research turned up some interesting information. For one thing, scientists learned that Americans were exercising less. By the 1920s, more people were starting to own cars. Over time, cities became more spread out. New areas were designed for driving, not for walking. As a result, people started getting less physical activity.

Americans also began to eat more processed foods. Today, fast foods and packaged snacks are everywhere. These foods contain far more sugar, salt, calories, and fat than whole foods do. Eating lots of processed foods and greasy fast food makes it easy to gain weight. Weighing too much is stressful on the heart.

There is another danger of eating too much fat. It travels through the bloodstream. Over time, it starts to block the arteries. These tubes become narrow and hard. It becomes more difficult for blood to get from the heart to other parts of the body.

Today, the average American eats more and exercises less than ever before. Both behaviors help explain the large rise in heart disease.

Do you get enough activity to help keep your heart healthy?

We've come a long way since the 1920s. Many Americans still have heart problems. But doctors now know how to treat and help prevent heart disease. For one thing, everyone should add exercise into their daily routines. Doctors say kids should get an hour or more of activity every day. Soccer, swimming, walking, and biking are all good activities.

Certain foods can also help protect your heart and blood vessels. Heart-healthy foods include salmon, oatmeal, chickpeas, lentils, almonds, walnuts, tuna, tofu, brown rice, soymilk, spinach, blueberries, carrots, broccoli, and sweet potatoes. Which of your restaurant meals contain heart-healthy foods?

Heart disease also runs in families. Have any of your relatives had heart attacks? How much exercise do you get? Is your diet full of foods that are good for your heart or bad for your heart? How can you change your diet to be nicer to your heart?

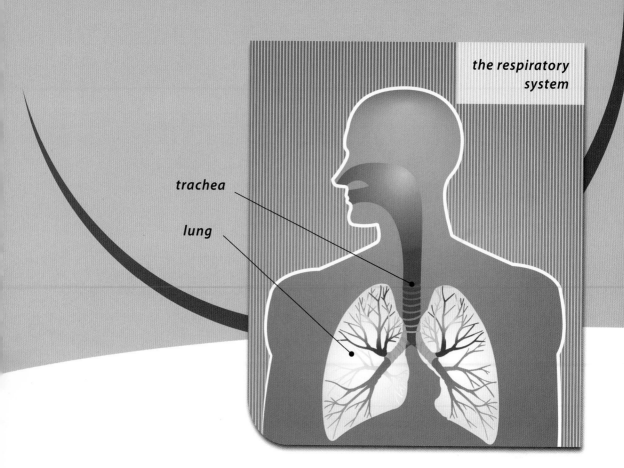

the respiratory system

trachea

lung

Keep Breathing

We need more than just food and water to survive and thrive. We also need oxygen. Oxygen is one gas in the air. When we inhale, we take it in through our trachea and into our lungs. When we exhale, we breathe out carbon dioxide and other gases.

Oxygen travels through the blood to every cell in the body. The gas helps our muscles turn food into fuel. Oxygen also aids digestion. How does your breathing change when you start running or biking? Why do you think your body might need to breathe harder or faster sometimes?

How Does the Nervous System Control Digestion?

You can digest food without thinking about it. Even without thinking, though, your brain plays an important role in digestion. So do your spinal cord and nerves. These are all part of your **nervous system.** The nervous system is what lets you taste, smell, and feel things, such as pain and hunger.

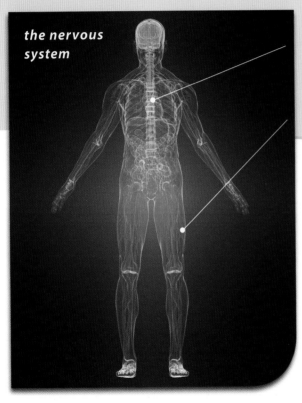

the nervous system

spinal cord

The red lines show nerves throughout the body.

Your nervous system directs the making of hunger hormones. Some of these hormones make you feel hungry and thirsty, so you'll eat and drink. Others make you feel full when it's time to stop. You will learn more about hormones on page 40.

Why is it a good idea to listen to your body for these signals? Do you think the runners will feel hungry during the race? What might be going on in their brains to distract them from wanting to eat?

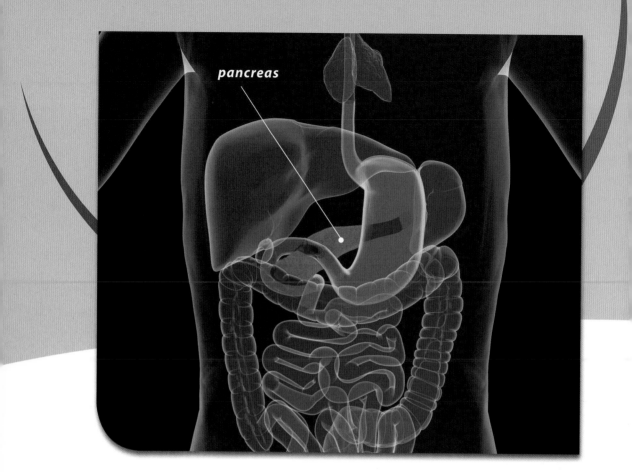

pancreas

The Helpers

Each system in your body helps the other systems. Even more help comes from the **endocrine glands.** These are structures located throughout your body. They make substances called hormones. Hormones act as chemical messengers. As they move through the bloodstream, they tell your cells and organs what to do. Hormones help people grow, deal with stress, and have babies.

The **pancreas** is the part of the endocrine system that is most helpful with digestion. You read about enzymes on page 30. Your pancreas makes these chemicals that help you break down food. This important gland also makes a hormone called **insulin** that regulates the amount of sugar that circulates in your blood.

Working It Out

When your body is done digesting your food, there is always some leftover waste. The **excretory system** moves solid and fluid wastes from your body. You have already learned that solid food wastes travel through the large intestine and then leave your body. Your kidneys do a lot of the work to rid the body of wastes in liquid form. All of the blood in your body goes through the kidneys. The equivalent of 400 gallons of blood flow through your kidneys each day. That would fill about eight average bath tubs!

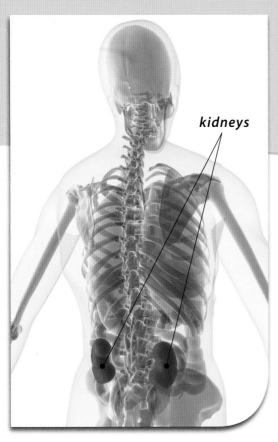

kidneys

When blood passes through them, the kidneys filter out extra vitamins, minerals and waste products that your body cannot use. You get rid of the waste, as well as excess water, when you use the bathroom.

As the runners get going, blood will start pumping faster through their bodies. What do you think that will do to their kidneys?

Now that you know how your body turns food to fuel, think again about how best to feed the runners at your restaurant. Let's consider the pros and cons of each meal:

Meal 1: A huge salad full of vegetables and fruit, a chocolate cookie, and a big glass of water.

Pros: Colorful fruits and vegetables contain lots of vitamins and minerals. Water will keep the runners hydrated.
Cons: There are too many simple sugars and not enough complex carbohydrates to keep the runners going over a long distance. Even though there is a lot of food here, there may not be enough calories in this meal.

Meal 2: A hamburger with lettuce, tomato, and ketchup; a handful of almonds; and a large soda.

Pros: Nuts contain healthy fats that are good for the heart. Meat contains iron, which is good for the blood, and protein.
Cons: There is too much protein in this meal for the night before the race. It would be better as a training meal. And soda has lots of added sugar, which might interrupt sleep.

Meal 3: A big plate of pasta with tomato sauce, shredded cheese, a small salad with walnuts, and a big glass of low-fat milk.

Pros: Lots of good carbohydrates. A good balance of carbs in the pasta, protein in the cheese and milk, and good fats in the nuts. Milk and cheese contain calcium for strong bones. There are plenty of calories here to fuel a long run.
Cons: Not many!

Meal 4: Fried fish sticks with tartar sauce, potato chips, and a blue sports drink.

Pros: Fish contains protein and good fats.
Cons: Fried foods like fish sticks and potato chips are greasy and can weigh a runner down. Sports drinks contain added sugars and fake colors. They are better during a long race than before it.

As you can see, the best meal for the night before the marathon is Meal 3. After the race, you might want to offer the runners bagels, bananas, and yogurt to help them recover. Good luck to the runners. And congratulations on a successful restaurant!

This runner knows how to eat to keep his body performing at its best!

Even if you don't run marathons, it is important to take care of your body by eating foods that are good for you. Look at everything you ate in a typical day before you read this book. In your food journal, continue to write down what you eat for a week. Compare your diet today with your diet a week ago and a week from now. Are you eating a good balance of carbohydrates, proteins, and fats? Are you getting enough vitamins and minerals? Are you eating too many calories? Are you eating too few calories? The best diet contains neither too much nor too little nutrition. Find *your* balanced diet.

It takes work every day to pick the best foods. And nobody is perfect. Keep working at it. Your body will thank you for the effort.

amino acids: basic building blocks of protein.

calories: units of measurement of the amount of energy in food.

carbohydrates: the sugars and starches that form the main source of energy in food.

circulatory system: the heart, blood, and blood vessels.

complete proteins: foods that have the correct balance of chemicals to support humans and other animals.

dehydration: extreme lack of water in the body.

endocrine glands: organs throughout your body that make hormones.

enzymes: protein substances that speed up the digestive process.

esophagus: tube that transports food from the mouth to the stomach.

excretory system: the system that rids the body of its waste products.

fats: nutrients our bodies need for energy and that help some vitamins do their jobs.

fiber: plant materials that cannot be digested, but that are necessary to keep the body working properly.

incomplete proteins: foods that do not have the correct balance of chemicals to support humans and other animals.

insulin: a hormone that manages the level of sugar in the blood.

metabolism: chemical process in the body that turns food into energy.

minerals: elements the body needs for growth and repair.

molecules: a group of two or more atoms that are bonded together.

mucus: a slippery substance made by the body's mucous membranes.

nervous system: the control center for the body, made up of the brain, spinal cord, and nerves.

nutrients: the substances found in food that are necessary for life.

nutrition: the process by which a living thing gets food and uses it.

pancreas: a gland that produces digestive juices, as well as the hormone insulin.

processed: altered in some way, often to make foods last longer on the grocery shelf.

proteins: nutrients that build and maintain the body.

respiration: the process in which oxygen is transported to cells.

unprocessed: still in a natural state, such as a piece of whole fruit.

vitamins: organic substances that the body needs for good health.

FURTHER READING

Food for Thought by Ken Robbins. Roaring Brook Press, 2009.

101 Things Everyone Should Know About Science by Dia Michels and Nathan Levy. Science Naturally!, 2009.

Wonderwise: Women in Science Learning Series. http://wonderwise.unl.edu

Amazing Food Detective.
http://members.kaiserpermanente.org/kpweb/richmedia/feature/amazingFoodDetective/index.htm

Food Guide Pyramid. http://www.mypyramid.gov/kids/index.html

ADDITIONAL NOTES

The page references below provide answers to questions asked throughout the book. Questions whose answers will vary are not addressed.

Page 8: You should make oils the smallest part of your food intake. The food pyramid can help you figure out how much food from each group to eat.

Page 9: Meal 3, followed by Meals 1 and 2. Meal 4 has the fewest colors.

Page 10: Meal 1; the vegetables and fruits in the salad get their energy directly from the Sun, as do the lettuce, tomato, and almonds in Meal 2. Meal 3; the salad and walnuts get their energy directly from the Sun. Meal 4; the potatoes in the chips got their energy directly from the Sun.

Page 11: Probably the cookie in Meal 1, the hamburger and soda in Meal 2, the cheese in Meal 3, and the fried fish and the chips in Meal 4 have the most calories. The glass of water in Meal 1 has no calories, and the lettuce in Meal 2 has few calories. A big runner will probably need more calories than a small runner, because a big runner has more body mass to consume fuel.

Page 12: Too few calories would provide too little fuel. Too many calories might slow a runner down, as the body spent energy in digesting the excess food. A runner should consume extra calories the night before a race. Calories are important, but an athlete runs faster and farther because of training and proper nutrition.

Page 14: The hamburger and the fish have the most protein. The chips and the fried fish have the most fat. The cookie probably has the most carbohydrates.

Page 15: The pasta has the most starch and the cookie has the most sugar. Starch would provide a longer-lasting fuel for runners.

Page 16: Empty calories come from foods that lack nutrition. Only the fruits, vegetables, and water are unprocessed. Any food that provides nutrition will help.

Page 17: Soda has the most sugar. Milk and water are best; soda and sports drinks are the worst.

Page 18: A few weeks before the race, runners should eat healthy foods. Salads, nuts, and milk are good choices. Hamburger and fish would provide protein to build muscles. The day before the race, pasta would provide fuel. The day after, foods high in proteins and carbohydrates would repair muscles and replace fuel.

Page 19: The pasta and the salad are vegetarian choices.

Page 20: Meal 2 and Meal 4 probably have the right amount of protein. Meal 1 has too little protein. Meal 3 may not have quite enough.

Page 21: Nuts: good fats; hamburger, cheese, and fried fish: bad fats

Page 22: Substitute healthy fats or cut down on fats in general.

Page 23: The runners should get enough of all the vitamins.

Page 24: milk, cheese have calcium; lettuce has potassium; hamburger has iron.

Page 25: Water. Runners should drink water to stay hydrated.

Page 26: Runners should eat slowly.

Page 27: Sharp teeth are good for biting and wide teeth are good for grinding.

Page 29: The hamburger and fish would be easiest to chew, and the almonds and walnuts would be harder to grind down.

Page 30: The runner should eat breakfast no later than 6:00 a.m.

Page 31: foods that are easy to digest so it's not sitting in their stomachs during a race. Also, they want the energy from the food available during the race.

Page 34: Iron is good for the blood. Leafy greens, red meats, beans, and breakfast cereals contain iron. Caption question: The heart is the red structure in the chest.

Page 37: Meal 2 contains heart-healthy almonds and Meal 3 has almonds.

Page 38: Breathing gets harder and faster during exercise. When the body is working harder, it needs more oxygen to fuel the cells.

Page 39: If you pay attention to the signals your body is sending, you will eat and drink enough, but not too much. Runners will probably not feel hungry, because the body will be too busy to do the work of digesting food.

Page 41: It will make their kidneys work harder.

INDEX